Newton's Laws of Motion

Preface

Newton's laws of motion are fundamental to understanding how the forces that act on objects compel them to move. While the laws appear simple and obvious, they present very elegant description on the interaction between force and matter. The laws have witnessed ground-breaking applications in the dynamics of objects like Earth satellites and planets.

Our thinking is that it is very important for science students to understand fundamental truths about these laws, and so we have developed this book with this in mind. The book presents very clear explanations of things we think students ought to know about the Newton laws. There are also numerical examples to guide the students' understanding quantitatively, and exercises to test their understanding of the topic.

NEWTON'S LAWS OF MOTION

Sir Isaac Newton (1642-1727) stated the three simple but important laws called "Newton's laws of motion."

Newton's laws of motion are theories of motion based on the idea of mass and force and the law connecting these physical concepts to the kinematic quantities such as displacement, velocity and acceleration.

Since the force responsible for the motion is taken into consideration in this case, we are discussing the dynamics of bodies in motion.

1. Newton's first law

1

This law states that: a body will continue in its state of rest, or if it is in motion, will continue to move with uniform speed in a straight line unless it is acted upon by a net external force.

Inertia

2

This tendency of a body to remain in its state of rest or in uniform motion is called inertia of the body. For this reason Newton's first law is sometimes called 'the law of inertia'. It is a measure of the mass of a body.

Consequences

3

An example of the consequences of the first law is when a moving vehicle is suddenly brought to rest by the application of the brakes. The passengers are suddenly jerked forward as they tend to continue in their straight line of motion.

Jamb 1992

4

A boy sits in a train moving with uniform speed on a straight track. If from his outstretched palm he gently tosses a coin vertically upwards, the coin will fall (A) In front of his palm   (B) beside his palm   (C) behind his palm

(D) Into his palm.

Answer (D) Into his palm

How?

5

Since the train is moving with uniform velocity in a straight track. And so long as the out stretched palm is within the train, the coin will fall into his palm. This is because the train and the coin have the same velocity. (I.e. will cover equal distances in equal time interval).

2. Momentum

6

The term momentum is also used in ordinary social discuss to mean different things. In physics momentum is an important property of a moving object. It explains the tendency of a body to continue moving in a straight line.

Definition

7

Momentum of a body can be understood as the quantity of motion in a body. But more formerly, it is defined as the product of mass and velocity of a body. i.e. momentum = M x V.

The unit of momentum is $kgms^{-1}$. In S.I units it is more commonly expressed as NS.

3. Newton's Second Law

8

This law states that the rate of change of momentum of a body is directly proportional to the net force applied and takes place in the direction in which the net force acts. Mathematically:

$$Force_{net} \propto \frac{Change\ in\ momentum}{time}$$

Why net force?

We say net force because, if the forces involved are not one, then we are only interested in the resultant force which is termed 'net' force.

Definition of force

Suppose a force f acts on a body of mass m for a time t and causes its velocity to change from u to v, then Newton's second law can be stated as:

$$f \propto \frac{mv - mu}{t}$$

$$f \propto \frac{m(v - u)}{t}$$

$$but \quad \frac{v - u}{t} = a$$

$$\therefore \quad f \propto m\,a$$

$$\Rightarrow f = k\,m\,a$$

where k is a constant.

Plan 11

If we take m = 1kg and a = 1m/s^2, the unit of force is chosen to make f = 1 when k = 1.

The S.I unit of force is called the Newton (N). The Newton (N) is the force which produces an acceleration of 1m/s^2 when it acts on a mass of 1kg.

Therefore, when f is in Newton (N), m in kg, and a in ms^{-2}, we have that:

f = ma --1

12

A body of mass 2kg moving vertically upwards has its velocity increased uniformly from 10ms^{-1} to 40ms^{-1} in 4s. Neglecting air resistance, calculate the upward vertical force acting on the body.

(A) 15N (B) 20N (C) 135N (D) 45N

Answer (A) 15N

How?

13

Solution:

From second law of motion:

$$f = ma = \frac{mv - mu}{t}$$

$$\therefore f = \frac{m(v - u)}{t}$$

but m = 2kg, v = 40ms^{-1}, u = 10ms^{-1} and t = 4s

$$f = \frac{2(40 - 10)}{4} = \frac{2(30)}{4} = 15N$$

Impulse of a force

14

A very important quantity in physics called 'impulse of a force' can be deduced from Newton's second law as follows:

$$f = \frac{m(v-u)}{t}$$

Multiplying both sides of the equation by t gives:

$$ft = m(v-u).$$

Plan 15

15

The quantity 'ft' in the above equation is called the impulse of the force I.

The unit of impulse is the Newton second (Ns).

Since this quantity is equal to change in momentum, it means that momentum can also be expressed in Ns.

i.e. ft = I = change in momentum.

Jamb 1998

16

The physical quantity that has the same dimensions as impulse is

(A) Energy   (B) Momentum   (C) Surface tension (D) Pressure

Answer (B) Momentum

| 17 |

A body of mass 100g moving with a velocity of 10.0ms^{-1} collides with a wall; if after the collision, it moves with a velocity of 2.0ms^{-1} in the opposite direction calculate the change in momentum.

(A) 0.8 Ns   (B) 1.2 Ns   (C) 12 Ns   (D) 80 Ns

Answer (B) 1.2 Ns

How?

| 18 |

I = change in momentum = m (v – u)

but m = 0.1kg, u = 10ms^{-1}, v = -2ms^{-1} (the negative sign indicates opposite direction)

∴ I = 0.1 [10 – (– 2)]

   = 0.1 (12)

   = 1.2 Ns

4. Force due to gravity – weight

| 19 |

The force of gravitational attraction of the earth on an object is called the weight w of the object.

The Newton's second law for a body undergoing a free fall in a gravitational field is given by w = mg, where w = f is the gravitational force

and a = g is the acceleration due to gravity.

Weight w = mg -- 2

20

From equation 2 above it can be concluded that the weight of an object is proportional to its mass, since g is the same for all objects at a given point.

The vector g is the force per unit mass exerted by the earth on any object. It is called the gravitational field of the earth.

Is g the same everywhere?

21

The answer is No!

Careful measurements of g at various places show that it does not have the same value everywhere. For example, its value on the earth is not the same with that on the moon.

It also varies with altitude; it varies inversely with the square of the distance of an object from the centre of the earth. Thus an object weighs slightly less at very high altitudes than it does at sea level.

Is there any difference between mass and weight?

22

Yes!

Mass	**Weight**
The mass of a body is the quantity of matter in the body	Weight w, of a body is the force of gravity on the body.
Same everywhere	Varies from place to place especially in different gravitational fields

5. Newton's third law

23

Newton's third law is sometimes called the law of interaction. This law states that: to every action, there is an equal and opposite reaction. It describes an important property of forces that they always occur in pairs.

Examples

24

When we place an object on a table, the reaction of the table on the objects is equal and opposite to the action of the object on the table.

Again, if a moving car A, hits a stationary car B, the force exerted on B by A will be the same as the action of B on A. The effect is that both cars would damage.

Weight of a body in a lift

25

A lift is a device that operates electrically and used for moving people and load up and down a tall building.

For a man standing on the lift, there are two forces acting on him. Namely:

(a) His weight (w) acting downwards and

(b) The normal Reaction (R) of the floor of the lift on the man acting upward.

Now we will now discuss the interaction of this forces when

    (i)   The lift is stationary

    (ii)   The lift accelerates upwards

    (iii)   The lift accelerates downwards with (a)

    (iv)   The lift descends with an acceleration $a = g$ (free fall).

A man on a stationary lift

26

When the lift is stationary or moving with a constant velocity, we have that

w = mg = R.

where m is the mass of the man

i.e. the weight of the man is equal to the reaction of the floor of the lift on the man.

A lift accelerating upwards

27

When the lift accelerates upwards with an acceleration, a, the man is pulled upwards with an acceleration, a; then the unbalanced force on him is given by:

F = R – mg = ma ---------------------------------- 3

Hence R = m (a + g)

Implication

28

Apparent weight of the man when the lift accelerates upwards is given by:

W = R = m (a + g) -------------------------------- 4

The implication is that the man appears to weigh more under this condition.

A lift accelerating downwards

29

When the lift moves downwards with an acceleration a, the unbalanced force (F) on the man is given by:

$F = mg - R = ma$.

His apparent weight is now $W = R = mg - ma$

or $W = m(g - a)$ -------------------------------------- 5

Thus the man appears to weigh less under this condition.

Weightlessness

30

When the lift descends with an acceleration $a = g$ then we can see from equation (5) above that $W = 0$.

Thus the man's apparent weight is zero.

Such a situation is referred to as "weightlessness".

Jamb 2006

31

A body weighing 80N stands in an elevator that is about to move. The force exerted by the floor on the body as the elevator moves upwards with an acceleration of $5ms^{-2}$ is

(A) 40N (B) 80N (C) 120N (D) 160N

$[g = 10s^{-2}]$

Answer (C) 120N

32

$W = 80N$, $a = 5ms^{-2}$, $g = 10ms^{-2}$

Since $W = mg$, the mass of the body is $m = W/g = 80/10 = 8kg$

Now, from equation 3, the unbalance force on the body is given by:

$F = R-W = ma$

$\therefore R = ma + W$

$= 8 \times 5 + 80$

$= 40 + 80 \quad = 120N$

Jamb 1995

33

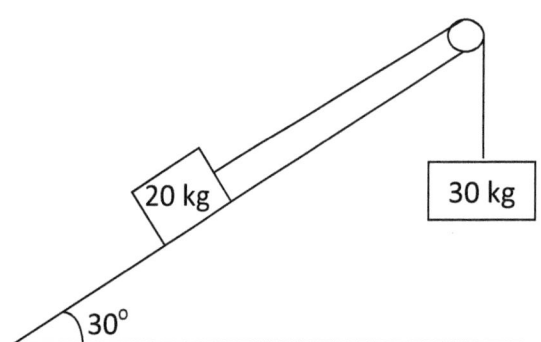

The acceleration of the system shown above is

(A) $2ms^{-2}$ (B) $4ms^{-2}$ (C) $6ms^{-2}$ (D) $8ms^{-2}$

Answer: (B) $4ms^{-2}$

How?

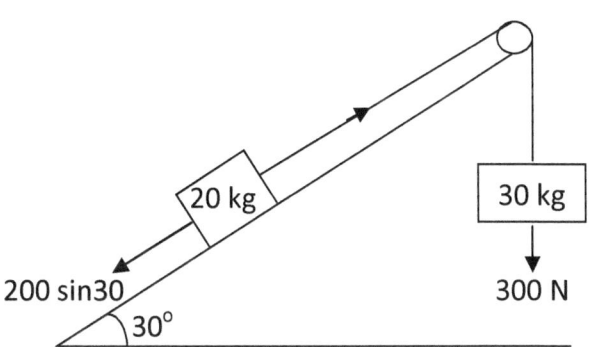

Force pulling the 20 kg mass down the plane = 200 sin30 = 200 × 0.5 = 100 N

Force pulling it up up the plane = weight of 30 kg mass = 300 N

Therefore, net force pulling it up the plane = 300 – 100 = 200 N

Now F = ma = 200N

Where m = (20 + 30) = 50 kg

$$\therefore a = \frac{F}{m} = \frac{200}{50} = 4ms^{-2}$$

6. Law of conservation of momentum

Newton's second and third laws enable us to formulate an important conservation law known as the law of conservation of momentum.

It states that: the total momentum of an isolated system of colliding bodies remains constant.

What do we mean by an isolated system?

36

By an isolated system, we mean that system in which no external force acts.

Let U_1 and U_2 and V_1 and V_2 be the initial and final velocities of two colliding bodies of masses M1 and M2. The conservation law can now be stated as:

$$M_1U_1 + M_2U_2 = M_1V_1 + M_2V_2$$

Note!

37

All the velocities must be measured in the same direction along the same line, with correct positive or negative signs as we illustrate shortly.

Plan 38

38

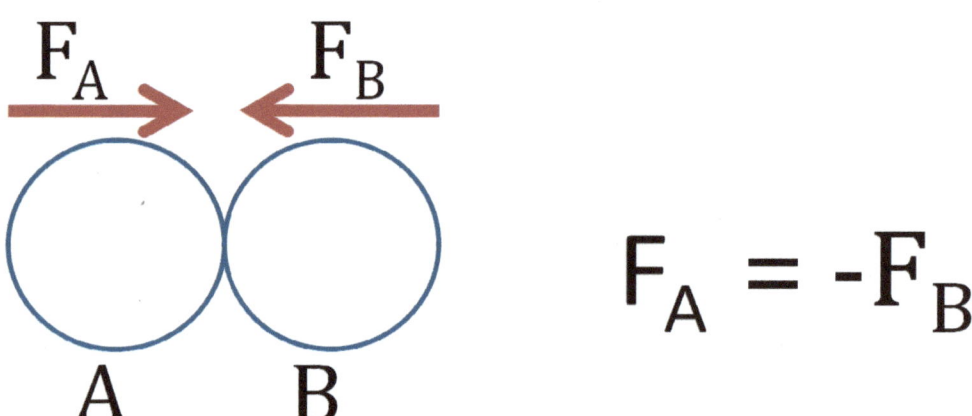

$$F_A = -F_B$$

Fig 1. Conservation of momentum

The above law follows from the fact that the action of a body 'A' on body 'B' is equal and opposite to the reaction of body 'B' on body 'A', and that both forces act for exactly equal times.

Cases

This argument applies to bodies which are elastic and rebound from each other after collision with different velocities (case 1).

It also applies to bodies which are inelastic and join together after the collision and move away with the same velocity (case 2).

In case 2: $V_1 = V_2 = V$

Therefore we can write:
$M_1U_1 + M_2U_2 = (M_1 + M_2) V$

Elastic collision (case 1)

In an elastic collision, both momentum and kinetic energy are conserved.

This means that for two colliding bodies with masses M_1 and M_2 and initial velocities of U_1 and U_2 and final velocities after collision of V_1 and V_2. Then
$M_1U_1 + M_2U_2 = M_1V_1 + M_2V_2$

$$\frac{1}{2}M_1U_1^2 + \frac{1}{2}M_2U_2^2 = \frac{1}{2}M_1V_1^2 + \frac{1}{2}M_2V_2^2$$

An example of perfectly elastic collision is a ball which bounces on the ground back to its original height.

In elastic collision (case 2)

41

In this case momentum is conserved but not the kinetic energy. The kinetic energy usually, decreases as it is converted into sound or elastic potential energy and in this way, causes deformation.

Go on

42

In a complete inelastic collision the two objects join together after an impact and move with the same velocity V, so that conservation of momentum equation becomes:

$M_1U_1 + M_2U_2 = (M_1 + M_2)V$

The K.E of the system before impact is

$K_1 = \frac{1}{2}M_1U_1^2 + \frac{1}{2}M_2U_2^2$

The K.E after impact is

$K_2 = \frac{1}{2}(M_1 + M_2)V^2$

Another Case

43

Consider the case when the body M_2 is at rest then,

$K_1 = \frac{1}{2}M_1U_1^2,$ $\qquad K_2 = \frac{1}{2}(M_1 + M_2)V^2$

But $V = \dfrac{M_1U_1}{M_2 + M_2}$ $\qquad \therefore K_2 = \frac{1}{2}\dfrac{M_1^2U_1^2}{M_1 + M_2}$

Then $\dfrac{K_1}{K_2} = \dfrac{M_1 + M_2}{M_1}$

This shows that the final kinetic energy K_2 of the body is less than the initial kinetic energy K_1.

| **44** |

An arrow of mass 0.1kg moving with a horizontal velocity of 15ms^{-1} is shot into a wooden block of mass 0.4kg lying at rest on a smooth horizontal surface. Their common velocity after impact is

(A)15ms^{-1} (B) 7.5ms^{-1} (C) 3.8ms^{-1} (D) 3.0ms^{-1}

Answer (D) 3.0ms^{-1}

How?

| **45** |

$M_1 = 0.1$kg, $U_1 = 15$ms^{-1}, $M_2 = 0.4$kg, $U_2 = 0$

But $M_1U_1 + M_2U_2 = (M_1 + M_2)$ V,   V =?

$0.1 \times 15 + 0 = (0.1 + 0.4)$V

$\therefore$ V = $\dfrac{1.5}{0.5}$ = 3.0ms^{-1}

| **46** |

A rocket burns fuel at the rate of 10kgs^{-1} and ejects it with a velocity of 5$\times$ 10^{3}ms^{-1}. the thrust exerted by the gas on the rocket is

(A) 2.5 $\times$ 10^{7}N   (B) 5 $\times$10^{4}N (C) 5$\times$ 10^{3}N (D) 2$\times$ 10^{2}N

Answer (B) 5 $\times$ 10^{4}N

How?

The thrust exerted by the gas on the rocket is equal and opposite of the force (action) provided by the engine in burning the fuel.

i.e. Thrust = action provided by engine.

$$= \text{mass per second} \times \text{velocity}$$

$$= 10 \text{kgs}^{-1} \times 5 \times 10^3 \text{ms}^{-1}$$

$$= 5.0 \times 10^4 \text{N}$$

WAEC 2000

A ball P of mass 0.25kg, losses one-third of its velocity when it makes a head on collision with an identical ball Q at rest. After the collision, Q moves off with a speed of 2ms^{-1} in the original direction of P. Calculate the initial velocity of P.

Plan 49

Solution

$M_1 = M_P = 0.25\text{kg}$, $M_2 = M_Q = 0.25\text{kg}$

$U_1 = U_P = ?$, $U_2 = U_Q = 0$

$V_P = \frac{1}{3}U_P = V_1$, $V_Q = V_2 = 2\text{ms}^{-1}$

By the law of conservation of momentum , $M_1 U_1 + M_2 U_2 = M_1 V_1 + M_2 V_2$

$$\Rightarrow M_P U_P + M_Q U_Q = \frac{1}{3}M_P U_P + M_Q V_Q$$

$$= 0.25 U_P + 0.25 \times 0 = \frac{1}{3}0.25 U_P + 0.25 \times 2$$

$0.2U_P - \frac{1}{3}0.25U_P = 0.5$

$0.17U_P = 0.5, \quad U_P = \dfrac{0.5}{0.17} = 3ms^{-1}$

(1)   A ball of mass 0.1kg is kicked against a rigid vertical wall with a horizontal velocity of 40ms^{-1}. If it rebounded with a horizontal velocity of 20m/s, calculate the impulse of the ball on the wall.

(A) 6.0Ns (B) 4.5Ns (C) 7.5Ns (D) 12Ns

(2) In an experiment, a force of 40N is applied to a mass of 10kg and the corresponding acceleration (a) is measured. Calculate the value of the acceleration.

(A) 4.5 ms^{-2} (B) 4.0 ms^{-2} (C) 3.5 ms^{-2} (D) 3.0 ms^{-2}

(3) When taking a penalty kick, a footballer applies a force of 30.0N for a period of 0.05s. If the mass of the ball is 0.075kg, calculate the speed with which the football moves off.

(A) 4.5ms^{-1} (B) 40.0ms^{-1} (C) 20.0ms^{-1} (D) 45.0ms^{-1}

(4) Two bodies have masses in the ratio 3:1. They experience forces which impact to them accelerations in the ratio 2:9 respectively. Find the ratio of the forces the masses experience.

(A) 1:4 (B)2:1 (C) 2:3 (D) 2:5

(Jamb 1999)

(5) A lead bullet of mass 0.05kg is fired with a velocity of 200ms^{-1} into a lead block of mass 0.95kg. Given that the lead block can move freely, the final kinetic energy after impact is

(A) 50J (B) 100J (C) 150J (D) 200J.

(Jamb 1999)

(6) A body of mass 4kg is acted on by a constant force of 12N for 3seconds. The kinetic energy gained by the body at the end of the time is

(A) 144J   (B) 162J   (C) 81J   (D) 72J

(7) A force of 100N is used to kick a football of mass 0.8kg. Find the velocity with which the ball moves if it takes 0.8seconds to be kicked. (A) $32ms^{-1}$ (B) $50ms^{-1}$ (C) $64ms^{-1}$ (D) $100ms^{-1}$.

(Jamb 2003)

(8) A rope is being used to pull a mass of 10kg vertically upward. Determine the fusion in the rope if, starting from rest, the mass acquire a velocity of $4ms^{-1}$ in 8seconds. (A) 5N   (B) 50N   (C) 95N   (D) 105N.

Solutions to the Exercise

1. (A)

2. (B)

3. (C)

4. (C)

5. (A)

6. (B)

7. (D)

8. (A)

www.ingramcontent.com/pod-product-compliance
Lightning Source LLC
Chambersburg PA
CBHW050435180526
45159CB00006B/2552